BEEKEEPING FOR BEGINNERS

A Comprehensive Guide to Sustainable Bee Farming, Hive Management, Honey Production, Breeding and Raising Queens

Anthony S. Cooper

Reserved rights apply. Without the publisher's prior written consent, no portion of this publication may be copied, distributed, or transmitted in any way, including by photocopying, recording, or other mechanical or electronic means, with the exception of brief quotations used in critical reviews and other noncommercial uses allowed by copyright law.

Copyright © Anthony S. Cooper, 2024.

INTRODUCTION 6

CHAPTER 1 10

THE BASICS OF BEEKEEPING 10

Understanding Bee Biology and Behavior 10

Types of Bees in the Hive: Workers, Drones, and Queens 14

Basic Beekeeping Terminology 18

CHAPTER 2 25

PLANNING YOUR SUSTAINABLE APIARY 25

Site Selection for Your Apiary 25

Designing an Eco-Friendly Apiary Layout 31

Choosing Sustainable Beekeeping Equipment 35

CHAPTER 3 41

STARTING YOUR SUSTAINABLE APIARY 41

Acquiring Bees: Packages, Nucs, and Swarms 41

Installing and Setting Up Your First Hives 45

Initial Hive Inspections and Best Practices 49

CHAPTER 4 55

HIVE MANAGEMENT FOR SUSTAINABILITY 55

Feeding Bees Naturally 55

Seasonal Hive Management Practices 59

Sustainable Hive Inspection Techniques 63

CHAPTER 5 67

BEE HEALTH AND WELLNESS 67

Identifying and Preventing Common Bee Diseases 67

Natural Remedies and Treatments for Bee Health 70

Monitoring and Maintaining Hive Wellness 76

CHAPTER 6 81

ENHANCING HONEY PRODUCTION SUSTAINABLY 81

Understanding Honey Flow and Forage 81

Techniques to Boost Honey Yield Naturally 85

Harvesting Honey with Minimal Impact 93

CHAPTER 7 98

PROCESSING AND SELLING HONEY 98

Sustainable Honey Processing Methods 98

Packaging and Labeling for Eco-Friendly Sales 104

CHAPTER 8 112

BREEDING AND RAISING QUEENS 112

Sustainable Queen Rearing Practices 112

Selecting and Breeding for Desirable Traits 117

Integrating New Queens into Your Apiary 123

CHAPTER 9 130

EXPANDING AND DIVERSIFYING YOUR APIARY 130

Increasing Hive Numbers Sustainably 130

Exploring Alternative Apiary Products (beeswax, propolis, etc.) 133

Value-Added Products and Services 136

CHAPTER 10 140

INTEGRATING BEEKEEPING WITH LOCAL ECOSYSTEMS 140

Enhancing Biodiversity in Your Apiary 140

Planting for Pollinators: Creating Bee-Friendly Habitats 142

Working with Local Farmers and Gardeners 145

CONCLUSION 148

INTRODUCTION

Sustainable beekeeping isn't just a buzzword; it's a fundamental approach to ensuring the survival and well-being of honeybee populations and the ecosystems they support. In recent years, the plight of bees has gained widespread attention due to factors such as habitat loss, pesticide exposure, climate change, and the spread of pests and diseases. As beekeepers, we play a crucial role in mitigating these threats and promoting the health and resilience of bee populations. Sustainable beekeeping practices prioritize the long-term health of bees and their environment over short-term gains. By adopting sustainable methods, we can minimize stress on bee colonies, reduce environmental impact, and foster biodiversity in our apiaries and surrounding landscapes. Sustainable beekeeping is not only about maximizing honey production but also about nurturing healthy colonies, preserving native habitats, and promoting pollinator health. Imagine an apiary where bees thrive in a diverse landscape of flowering plants, trees, and wild habitats. Picture hives buzzing with activity as bee's forage for nectar and pollen, pollinating crops and wildflowers alike. Envision beekeepers tending to their colonies with care and respect, employing practices that support bee health and ecological balance. In our vision for sustainable apiaries, beekeeping goes beyond mere honey production to become a harmonious partnership between humans and bees, rooted in principles of environmental stewardship and conservation. It's a vision where apiaries serve as sanctuaries for pollinators, havens of biodiversity, and centers of community engagement and education. By embracing this vision, we can create a world where bees thrive, ecosystems flourish, and sustainable agriculture becomes the norm.

This manual serves as a comprehensive guide to sustainable beekeeping, offering practical advice, insights, and strategies for beekeepers at all levels of experience. Whether you're a beginner just starting your journey in beekeeping or a seasoned apiarist looking to refine your practices, this manual is designed to provide you with the knowledge and tools you need to succeed. Each chapter explores a different aspect of sustainable beekeeping, from hive management and honey production to habitat enhancement and community outreach. Within each chapter, you'll find detailed explanations, step-by-step instructions, and practical tips to help you implement sustainable practices in your apiary. As you navigate this manual, we encourage you to approach beekeeping with curiosity, compassion, and a commitment to sustainability. Take the time to observe your bees, learn from their behavior, and adapt your practices to meet their needs. Remember that beekeeping is as much an art as it is a science, and there's always more to discover and explore.

CHAPTER 1

THE BASICS OF BEEKEEPING

Understanding Bee Biology and Behavior

The species Apis mellifera, commonly known as the Western honeybee, is the most frequently maintained by beekeepers due to its adaptability and prolific honey production. Originating from Europe, Africa, and parts of Asia, Apis mellifera has been spread worldwide due to human intervention. This species is known for its complex social behavior and efficient colony structure, making it an invaluable asset for both honey production and agricultural pollination. A honeybee's anatomy is highly specialized for its ecological roles. The head houses critical sensory organs, including compound eyes for detecting light and movement, antennae for sensing chemical signals, and mandibles for manipulating wax and feeding. The proboscis, a specialized mouthpart, allows bees to extract nectar from flowers. The thorax contains the muscles necessary for flight, with three pairs of legs and two pairs of wings attached. Each leg is adapted for different tasks, such as cleaning, pollen collection, and manipulating the hive. The abdomen houses vital organs, including the digestive and reproductive systems, and contains the sting apparatus, which is primarily used for defense. The life cycle of a honeybee consists of four distinct stages: egg, larva, pupa, and adult. A queen bee lays eggs in individual cells within the hive. These eggs hatch into larvae after about three days. Worker bees feed the larvae a diet of royal jelly initially, followed by a mixture of honey and pollen. After about six days, the larvae are sealed in their cells to pupate. During the pupal stage, which lasts approximately 12 days, the larvae undergo metamorphosis, transforming into adult bees. The entire process from egg to adult takes about 21 days for worker bees, slightly longer for drones (male bees), and about 16 days for queens. Honeybees exhibit sophisticated communication methods and social structures. One primary mode of communication is through pheromones, which are chemical signals that convey information about the hive's status, such as the presence of a queen or the need for food. Another remarkable form of communication is the waggle dance, performed by forager bees to inform others about the location of food sources. This dance conveys information about the direction and distance to flowers with nectar and pollen. The social hierarchy within a hive includes a single queen, whose primary role is reproduction, thousands of worker bees who maintain the hive and forage for food, and drones, whose sole purpose is to mate with a queen. Foraging behavior in honeybees is a complex and vital activity for the survival of the colony. Worker bees leave the hive to collect nectar and pollen from flowers. Nectar is converted into honey, which serves as a food source, while pollen provides essential proteins and lipids. Some worker bees, known as scout bees, search for new foraging sites and communicate their findings to the colony through the waggle dance. This behavior ensures the hive has a constant supply of resources, and bees exhibit remarkable memory and navigation skills to locate and return to productive flowers. Understanding these aspects of bee biology and behavior is essential for beekeepers to manage their hives effectively and support the health of these crucial pollinators.

Types of Bees in the Hive: Workers, Drones, and Queens

In the complex social structure of a honeybee hive, three types of bees—workers, drones, and queens—each have specialized roles that are crucial for the hive's function and survival. Worker bees are the most numerous members of the hive and take on a variety of roles throughout their lives. Initially, they function as nurse bees, tending to the larvae by feeding them royal jelly and later a mixture of honey and pollen. As they age, they transition into other roles such as guard bees, where they protect the hive entrance from intruders, and eventually become foragers. Foragers venture out to collect nectar, pollen, water, and propolis, essential resources for the hive's sustenance. The lifespan of a worker bee varies depending on the season, typically lasting several weeks during the active summer months and several months if they emerge in late autumn. As they age, their duties shift, starting from hive maintenance and care of the brood to guarding and ultimately foraging. Physically, worker bees are equipped with specialized structures such as pollen baskets on their hind legs, wax glands on their abdomen for comb building, and a proboscis for nectar collection. Their stingers, barbed and designed for defense, can only be used once, resulting in the bee's death post-sting. Drone bees, the male members of the hive, have a singular primary function: to mate with a queen from another hive. They develop from unfertilized eggs, and their life cycle progresses from egg to larva to pupa, eventually emerging as adult drones. Drones do not possess stingers, nor do they partake in foraging or hive maintenance. Instead, their bodies are adapted for mating flights, with larger eyes for better vision to locate queens during these flights. On mating flights, drones gather at drone congregation areas and await the arrival of virgin queens. After mating, drones die shortly afterward, as the act is fatal for them. Those that do not mate may be expelled from the hive before winter, as they become a resource drain. The queen bee, central to the hive's reproduction and social structure, has the primary role of laying eggs. She can lay thousands of eggs per day during peak seasons, ensuring the colony's population remains robust. The queen also produces pheromones that regulate the hive's activities and maintain social order, preventing other females from developing reproductive capabilities. Queens develop from larvae chosen by the worker bees and are fed exclusively with royal jelly, a nutrient-rich substance that triggers their development into a reproductive bee. Once fully developed, a virgin queen embarks on mating flights, during which she mates with several drones, storing their sperm in her spermatheca for use throughout her life. A queen's lifespan can range from two to five years, although her egg-laying capacity diminishes with age, often leading to her replacement by the hive.

The life cycles and roles of worker bees, drones, and queens are intricately designed to ensure the hive's survival, growth, and efficiency. Each type of bee has physical and behavioral adaptations that enable them to fulfill their specific duties, contributing to the colony's complex and cooperative social structure.

Basic Beekeeping Terminology

An apiary is a managed collection of beehives where bees are kept for honey production, pollination services, or both. These apiaries can range from small backyard setups to large-scale commercial operations. They are strategically located near flowering plants to ensure a good supply of nectar and pollen for the bees.

1. **Hive Components:**

- **Hive Box:** The hive box serves as the main structure where bees build their comb and store honey, pollen, and brood. Different types include the Langstroth hive, which consists of stackable boxes with removable frames, the top-bar hive, where bees build their comb freely from bars placed across the top, and the Warre hive, which uses boxes stacked from the bottom up, encouraging natural comb-building.
- **Frames and Foundation:** Frames provide support for the bees' comb and make hive inspection and manipulation easier for beekeepers. Foundation sheets, typically made of beeswax or plastic, guide the bees in building straight comb and can be installed in frames to accelerate comb construction.

- **Supers:** Supers are additional boxes placed above the brood chamber to provide extra space for honey storage. They allow beekeepers to harvest surplus honey without disturbing the brood nest.

- **Queen Excluder:** A queen excluder is a device placed between the brood chamber and honey supers to prevent the queen from laying eggs in the honey storage area. This ensures that the honey harvested is free from brood.

2. **Common Tools:**

- **Hive Tool:** A hive tool is a multipurpose tool used by beekeepers for various tasks such as prying apart hive components, scraping off propolis and excess wax, and lifting frames during inspections. Variants include J-hooks, scrapers, and frame lifters.

- **Smoker:** A smoker is a device that produces smoke used to calm bees during hive inspections. By puffing smoke into the hive, beekeepers disrupt the bees' communication and trigger a feeding response, causing them to consume honey and become less aggressive.
- **Bee Suit:** Beekeepers wear protective clothing such as bee suits, veils, gloves, and boots to shield themselves from bee stings. Suits come in various styles, including full-body suits, jackets with veils, and ventilated suits, providing different levels of protection depending on the beekeeper's preferences and the intensity of the beekeeping activity.

3. **Beekeeping Processes:**
- **Swarming:** Swarming occurs when a colony's population outgrows its space, prompting the

bees to form a new colony. Beekeepers manage swarming by providing adequate hive space, performing swarm prevention techniques, and capturing swarms to prevent them from establishing wild colonies.

- **Supering:** Supering involves adding supers to a hive when honey production increases or to prevent overcrowding during the nectar flow. This allows bees to store surplus honey without congesting the brood nest.
- **Extracting Honey:** Extracting honey involves removing frames of capped honey from the hive, uncapping the cells, and extracting the honey using centrifugal force or other methods. Beekeepers use extractors, uncapping knives, and strainers to harvest and process honey.

4. Health and Disease Terms:

- **Varroa Mite:** Varroa mites are external parasites that infest honeybee colonies, feeding on the bees' hemolymph (blood) and transmitting diseases. They weaken bees and can cause colony collapse if left untreated.

- **Foulbrood:** Foulbrood is a bacterial disease that affects bee brood, causing larvae to die and turn foul-smelling. It can be caused by two types of bacteria: American foulbrood (Paenibacillus larvae) and European foulbrood (Melissococcus plutonius). Treatment typically involves destroying infected hives and practicing good hygiene.

- **Nosema:** Nosema is a fungal infection caused by the microsporidian parasites Nosema apis and Nosema ceranae. It affects the bee's digestive system, leading to reduced lifespan, decreased foraging efficiency, and colony decline. Management strategies include medication and maintaining hive hygiene to reduce spore transmission.

CHAPTER 2

PLANNING YOUR SUSTAINABLE APIARY

Site Selection for Your Apiary

Site selection for an apiary is a crucial decision that significantly impacts the health, productivity, and sustainability of bee colonies. Understanding various environmental factors and considering urban versus rural settings are essential aspects of choosing the ideal location for your beekeeping operation.

1. **Understanding Environmental Factors:**

- **Sunlight:** Adequate sunlight is vital for hive health and productivity. Bees rely on sunlight for orientation, warmth, and activity regulation within the hive. Selecting a site with good sun exposure ensures that hives remain warm and dry, promoting brood rearing and honey production.

- **Wind Exposure:** Wind can affect hive stability and temperature regulation. Strong winds can topple hives and disrupt bees' ability to maintain optimal hive temperatures. Choosing a site with natural windbreaks, such as trees or buildings, helps protect hives from wind exposure and maintains a stable hive environment.

- **Water Sources:** Proximity to water sources is essential for bees' hydration needs. Bees require water for various hive activities, including cooling the hive, diluting honey, and feeding larvae. Selecting a site near reliable water sources such as streams, ponds, or birdbaths ensures that bees have easy access to water, reducing the risk of dehydration and improving hive health.

- **Forage Availability:** Assessing nearby flora for nectar and pollen sources is critical for beekeeping success. Diverse and abundant forage provides bees with a balanced diet and supports colony growth and honey production. Choose a location with a variety of flowering plants that bloom throughout the growing season to ensure continuous forage availability for your bees.

2. **Considerations for Urban vs. Rural Apiaries:**

- **Urban Challenges:** Beekeeping in urban environments presents unique challenges such as noise, pollution, and limited forage. Urban areas may have higher levels of noise pollution from traffic and human activities, which can stress bees and disrupt hive communication. Pollution from vehicles, industrial activities, and pesticide use can also negatively impact bee health. Additionally, urban environments often have limited green spaces and floral diversity, affecting bees' foraging opportunities.

- **Rural Advantages:** Rural areas offer advantages for beekeeping, including an abundance of natural forage and fewer environmental pollutants. Rural landscapes typically have larger green spaces, diverse flora, and fewer sources of pollution, providing bees with ample foraging opportunities and cleaner environments. Rural areas also tend to have fewer noise disturbances, allowing bees to focus on hive activities without interruptions.

- **Regulatory and Zoning Requirements:** Before establishing an apiary, it's essential to research local regulations regarding beekeeping, setbacks, and property zoning laws. Some municipalities have specific regulations governing beekeeping activities, such as hive placement, hive density per property, and registration requirements. Understanding and complying with these regulations ensures legal compliance and prevents potential conflicts with local authorities.

- **Neighbors and Community Relations:** Building positive relationships with neighbors and the community is crucial for successful beekeeping, particularly in urban areas. Communicate openly with neighbors about your beekeeping activities, addressing any concerns they may have regarding bee stings, swarming, or hive placement. Educating the community about the importance of bees and their role in pollination can foster support for your beekeeping endeavors and mitigate potential conflicts.

By carefully considering environmental factors, understanding the unique challenges and advantages of urban versus rural settings, and adhering to local regulations, you can select an optimal site for your apiary that promotes bee health, productivity, and positive community relations.

Designing an Eco-Friendly Apiary Layout

Maximizing biodiversity and minimizing environmental impact are crucial considerations for sustainable beekeeping practices. By incorporating native plantings, habitat enhancements, and minimizing chemical use, beekeepers can create thriving ecosystems that support healthy bee populations while minimizing harm to the environment.

1. **Maximizing Biodiversity:**

- **Native Plantings:** Incorporating native flowering plants in and around the apiary provides essential forage for pollinators, including honeybees. Native plants are well-adapted to local conditions and often require less maintenance than non-native species.

- **Wildflower Meadows:** Creating areas of diverse wildflowers, or wildflower meadows, can significantly increase the availability of nectar and pollen for foraging bees. These meadows provide a continuous and varied food source throughout the growing season, supporting bee health and productivity.
- **Habitat Enhancement:** Installing bee-friendly structures such as insect hotels, solitary bee nesting sites, and pollinator gardens enhances habitat diversity and provides nesting and foraging opportunities for native bees and other beneficial insects.

2. **Minimizing Environmental Impact:**
- **Chemical-Free Zones:** Avoiding pesticide and herbicide use near the apiary helps protect bees from exposure to harmful chemicals.

Instead, practice integrated pest management techniques and opt for organic and natural alternatives to control pests and weeds.

- **Water Conservation:** Implementing water-saving measures like rainwater harvesting reduces the demand for municipal water supplies and ensures a sustainable water source for bees. Providing clean water sources within the apiary also helps meet the bees' hydration needs.

- **Natural Pest Control:** Encouraging natural predators such as ladybugs, lacewings, and birds helps control pest populations in and around the apiary. Avoiding broad-spectrum pesticides preserves the natural balance of predator and prey species, reducing the need for chemical interventions.

3. Optimizing Hive Placement:

- **Spacing Between Hives:** Maintaining adequate distance between hives prevents overcrowding and minimizes drifting, where bees mistakenly enter the wrong hive. Proper hive spacing also facilitates hive management tasks and reduces the risk of disease transmission between colonies.

- **Hive Orientation:** Orienting hives to maximize sunlight exposure and minimize wind impact helps regulate hive temperature and promotes bee activity. Hives should face south or southeast to capture the warmth of the morning sun and be sheltered from prevailing winds.

- **Accessibility and Safety:** Ensure pathways and clearances around the apiary for beekeepers and visitors. Clear, well-maintained pathways and proper signage improve safety and accessibility, making hive inspections and maintenance tasks more manageable and reducing the risk of accidents.

Choosing Sustainable Beekeeping Equipment

Material selection is a critical aspect of beekeeping that can impact hive health, equipment longevity, and environmental sustainability. By choosing appropriate materials and adopting maintenance practices, beekeepers can ensure the longevity of their equipment while minimizing environmental impact.

1. **Material Selection:**
- **Wooden Equipment:** Natural materials like untreated wood are favored for hive bodies

and frames due to their breathability, insulation properties, and compatibility with beeswax. Wooden equipment provides a stable environment for bees, allowing moisture regulation and preventing condensation buildup inside the hive.

- **Avoiding Plastics:** Minimizing the use of plastic components in beehives is essential, as plastics can degrade over time, releasing harmful chemicals that can harm bees. While plastic frames and foundation may offer convenience, they may not provide the same level of durability and insulation as wooden alternatives.
- **Metal Components:** When selecting metal components such as hive hardware, prioritize durable, corrosion-resistant metals like

stainless steel or galvanized steel. These materials withstand the harsh conditions of the hive environment, ensuring longevity and reducing the need for frequent replacements.

2. **Equipment Lifespan and Maintenance:**

- **Longevity Considerations:** Investing in well-made, durable equipment may initially cost more but can significantly reduce waste and save money in the long run. High-quality wooden hives and metal hardware can withstand years of use with proper maintenance, ultimately minimizing the environmental impact of beekeeping operations.
- **Routine Maintenance:** Regular inspections and upkeep are essential to extend the lifespan of beekeeping equipment. Routine

maintenance tasks include cleaning hive components, replacing worn-out parts, and treating wooden surfaces to prevent rot and decay. By addressing issues promptly, beekeepers can prevent minor problems from escalating and prolong the lifespan of their equipment.

- **Repair and Reuse:** Whenever possible, prioritize repair and reuse over replacement. Damaged hive components can often be repaired with simple tools and techniques, extending their usability and reducing waste. By adopting a mindset of repair and reuse, beekeepers can minimize their environmental footprint and promote resource efficiency.

3. **Resource Efficiency:**

- **Space Optimization:** Choose hive designs that maximize colony health while minimizing material usage. Compact hive designs like the Warre hive or vertical stacking systems optimize space, allowing beekeepers to maintain healthy colonies without excessive material consumption.

- **Energy Efficiency:** Consider energy-efficient extraction methods and equipment for honey harvesting. Manual or low-energy extraction methods, such as crush-and-strain or hand-cranked extractors, reduce reliance on electricity and minimize environmental impact.

- **Transportation Impact:** Minimize the carbon footprint associated with beekeeping equipment by sourcing locally produced materials whenever possible. By supporting local suppliers and manufacturers, beekeepers can reduce transportation emissions and promote sustainable practices within their communities.

CHAPTER 3

STARTING YOUR SUSTAINABLE APIARY

Acquiring Bees: Packages, Nucs, and Swarms

Acquiring bees is a crucial step in starting or expanding a beekeeping operation, and understanding the different sources available—packages, nucleus colonies (nucs), and swarms—can help beekeepers make informed decisions that promote colony health and productivity.

1. **Understanding Bee Sources:**

- **Package Bees:** Package bees are typically purchased and consist of a screened box containing a queen and a set number of worker bees. These bees are often sourced from commercial bee breeders and are shipped in the mail. Package bees are a popular option for beekeepers looking to start new colonies or replace winter losses.

- **Nucleus Colonies (Nucs):** Nucleus colonies, or nucs, are small, established colonies containing a queen, frames of brood in various stages of development, honey stores, and a population of worker bees. Nucs provide a head start for beekeepers, as they already contain a functioning queen and brood, accelerating colony growth and honey production.
- **Capturing Swarms:** Swarms are groups of bees that have left their parent colony to establish a new hive. Beekeepers can capture swarms either from their own or neighboring colonies or receive swarms from local beekeeping associations. Capturing swarms is

an economical way to acquire bees, as they are typically free and already adapted to local conditions.

2. **Considerations for Bee Sources:**

- **Genetic Diversity:** When selecting bee sources, prioritize genetic diversity to promote colony resilience and productivity. Bees from diverse genetic backgrounds are better equipped to adapt to changing environmental conditions, resist pests and diseases, and forage efficiently.

- **Local Adaptation:** Preferentially source bees adapted to your region's climate and environmental conditions. Locally adapted bees have evolved traits that make them better suited to local flora, weather patterns, and environmental stressors, resulting in healthier and more productive colonies.

- **Disease Resistance:** Choose bee sources from suppliers with healthy colonies and good hygiene practices to minimize the risk of introducing diseases into your apiary. Inspect potential bee sources for signs of disease or pests and inquire about disease management practices, such as regular colony inspections and disease prevention measures.

By considering factors such as genetic diversity, local adaptation, and disease resistance when acquiring bees, beekeepers can establish strong and resilient colonies that thrive in their environment. Whether purchasing package bees, nucleus colonies, or capturing swarms, thoughtful selection of bee sources is essential for the long-term success of a beekeeping operation.

Installing and Setting Up Your First Hives

Preparing hive equipment, installing bees, and providing essential resources such as food and water are crucial steps in establishing healthy and thriving bee colonies. By carefully assembling and cleaning hive components, positioning hives appropriately, and following proper installation procedures, beekeepers can give their bees the best possible start. Before installing bees, assemble hive components such as hive bodies, frames, and supers, ensuring they are clean and free from contaminants. Cleanliness is essential to prevent the spread of diseases and pests within the hive. Scrub wooden components with a mild detergent and water solution, and sterilize metal hardware with a bleach solution or alcohol. Allow the equipment to dry completely before use. Position hives in the selected apiary site with careful consideration for sunlight exposure, wind protection, and accessibility. Orient hives to face south or southeast to maximize sunlight exposure and minimize wind impact. Ensure clear pathways and adequate space around the hives for beekeeper access and hive maintenance.

Installing Bees

- **Package Installation:** Installing packaged bees into their new hive involves several steps. Begin by removing the feeder can and queen cage from the package. Carefully shake the bees into the hive, ensuring the queen is safely released into the hive. Provide supplemental feeding with sugar syrup to help the bees establish themselves and draw comb. Close the hive and monitor the bees' progress in the coming days.

- **Nuc Installation:** Transferring frames from a nucleus colony (nuc) into the permanent hive is a straightforward process. Simply remove frames from the nuc box and transfer them into the corresponding slots in the hive body. Ensure the queen is safely transferred along with the brood frames. Provide supplemental feeding if necessary and close the hive.

- **Swarms:** Capturing and transferring swarms into hive equipment requires careful handling to avoid disturbing the bees. Begin by gently coaxing the swarm into a temporary container or swarm trap. Once the bees have settled, transfer them into the hive by shaking or brushing them in. Ensure the queen is safely transferred, as she is essential for the colony's survival. Close the hive and monitor the swarm's progress.

Feeding and Watering

- **Sugar Syrup:** Provide supplemental feeding with sugar syrup to newly installed colonies to help them establish and draw comb. Mix a solution of sugar and water at a ratio of 1:1 or 2:1, depending on the colony's needs and available forage. Place feeder jars or feeders inside the hive to allow bees easy access to the syrup.
- Water Sources: Ensure nearby water sources are available and accessible for newly installed

colonies. Bees require water for hydration and to regulate the temperature and humidity inside the hive. Place shallow dishes or water sources near the hives to provide bees with easy access to clean water. Monitor water levels regularly and refill as needed to ensure continuous access for the bees.

Initial Hive Inspections and Best Practices

Safety precautions are paramount in beekeeping to ensure the well-being of both the beekeeper and the bees. By wearing appropriate protective gear, conducting inspections in favorable weather conditions, and following proper inspection procedures, beekeepers can minimize risks and handle their colonies safely and effectively.

1. Safety Precautions:

- **Protective Gear:** Beekeepers should wear appropriate protective clothing and gear to minimize stings and injuries. This typically includes a bee suit or jacket, veil, gloves, and closed-toe shoes or boots. The gear should provide full coverage and protection against bee stings while allowing for flexibility and ease of movement during hive inspections.

- **Weather Considerations:** Conduct hive inspections during favorable weather conditions to avoid disturbing bees unnecessarily. Bees are more likely to be agitated during windy, rainy, or excessively hot weather, increasing the risk of defensive behavior and stings. Choose calm, mild days with moderate temperatures for hive inspections to minimize stress on the bees and ensure a safe working environment.

2. **Inspection Procedures:**
- **Observation Techniques:** Use smoke and gentle movements to calm bees during inspections. Smoking the hive disrupts communication among the bees and triggers a feeding response, making them less likely to sting. Move slowly and deliberately, avoiding sudden movements or loud noises that may agitate the bees. Maintain a relaxed demeanor and approach the hive with confidence to minimize disturbance.
- **Record-Keeping:** Document observations, hive conditions, and any actions taken during

inspections. Keeping detailed records allows beekeepers to track hive health and productivity over time, identify trends or patterns, and make informed management decisions. Note the date of the inspection, hive weight, population strength, brood patterns, honey stores, and any signs of pests or diseases.

3. **Identifying Healthy Colonies:**

- **Brood Patterns:** Assess the brood pattern for signs of health, including brood density, uniformity, and the presence of eggs. A healthy colony will exhibit a solid brood pattern with eggs, larvae, and capped brood in various stages of development. Irregular or spotty brood patterns may indicate issues such as queen problems, diseases, or nutritional deficiencies.

- **Population Strength:** Evaluate the size and activity level of the colony during inspections. A healthy colony will have a robust population of worker bees actively foraging, caring for brood, and maintaining hive cleanliness. Monitor population trends over time to detect changes or abnormalities that may signal health issues or environmental stressors.

4. Addressing Issues:

- **Pest and Disease Management:** Monitor for signs of pests and diseases during inspections and take proactive measures if detected. Conduct regular checks for common pests such as Varroa mites and inspect brood frames for signs of diseases such as American foulbrood or chalkbrood. Implement integrated pest management strategies to control pests and diseases while minimizing chemical interventions.

- **Feeding Considerations:** Adjust feeding strategies based on colony needs and nectar flow. Supplemental feeding may be necessary during periods of nectar dearth or when colonies are low on honey stores. Monitor hive weight and observe bee activity to determine feeding requirements and provide sugar syrup or other supplemental feed as needed to support colony health and survival.

CHAPTER 4

HIVE MANAGEMENT FOR SUSTAINABILITY

Feeding Bees Naturally

Feeding bees naturally involves understanding their nutritional needs, providing natural forage options, and employing feeding techniques that mimic their natural diet. By cultivating a diverse range of flowering plants, designing bee-friendly landscapes, and using supplemental feeding methods when necessary, beekeepers can support healthy and thriving bee colonies. Nectar serves as the primary source of carbohydrates for bees, providing them with energy for flight, hive maintenance, and brood rearing. Pollen, on the other hand, is rich in proteins, vitamins, and minerals essential for bee development and health. Bees collect pollen to feed developing larvae and sustain the adult population within the hive. Supplemental feeding may be necessary when natural forage options are limited, such as during times of nectar dearth, adverse weather conditions, or habitat loss. Providing supplemental feed ensures that bees have access to essential nutrients to maintain colony strength and vitality, especially during critical periods such as spring buildup or winter survival. Cultivating a diverse range of flowering plants in and around the apiary provides natural forage options for bees. Choose a variety of plant species that bloom at different times throughout the growing season to ensure continuous food sources for bees. Native wildflowers, herbs, fruit trees, and perennial flowers are excellent choices for supporting pollinator populations. Designing gardens and landscapes with bee-friendly plants not only enhances the aesthetic appeal but also supports pollinator health and biodiversity. Opt for plants that are rich in nectar and pollen, free from pesticides and herbicides, and suited to the local climate and soil conditions. Incorporate a mix of annuals, perennials, shrubs, and trees to provide diverse forage options for bees throughout the year.

Feeding Techniques

Sugar syrup is a common supplemental feed used by beekeepers to provide bees with carbohydrates during times of nectar dearth or when colonies require additional energy. Prepare sugar syrup by mixing granulated sugar and water at a ratio of 1:1 (by weight) or 2:1 (by volume) until the sugar is completely dissolved. Dispense the syrup using feeder jars, hive-top feeders, or frame feeders placed inside the hive for easy access by the bees. Commercial or homemade pollen substitutes can provide essential nutrients to colonies when natural pollen sources are scarce. These substitutes typically contain a mix of protein-rich ingredients such as soy flour, brewer's yeast, and dried milk powder. Follow manufacturer's instructions or recipes to prepare and dispense pollen substitutes, ensuring they are readily available to bees when needed. Feeding bees naturally involves providing a diverse and abundant forage environment while supplementing their diet when necessary to support colony health and productivity. By understanding bee nutritional needs and employing natural feeding alternatives and techniques, beekeepers can contribute to the well-being of their colonies and promote sustainable beekeeping practices.

Seasonal Hive Management Practices

Spring is a critical period for bee colonies as they prepare for the upcoming nectar flow. To stimulate brood rearing, beekeepers can provide supplemental feeding with sugar syrup to ensure the colony has ample energy reserves for raising brood. Additionally, techniques such as reversing hive bodies or checkerboarding frames can encourage the queen to lay eggs in the upper brood chambers, promoting rapid population growth. As colonies rapidly expand in spring, the risk of swarming increases. To mitigate this risk, beekeepers can implement swarm prevention strategies such as making splits to divide the colony, adding empty frames for expansion, and providing adequate ventilation to prevent congestion. Regular hive inspections allow beekeepers to identify signs of swarm preparations, such as queen cells, and take preemptive action to prevent swarming.

During the active foraging season, hive inspections should be conducted at regular intervals to monitor colony health and productivity. Inspections typically focus on assessing brood patterns, honey stores, and pest management. Beekeepers may choose to conduct weekly or biweekly inspections, depending on the colony's strength and forage availability. As nectar flow increases in summer, beekeepers add honey supers to accommodate surplus nectar collection. Supering involves placing additional boxes above the brood chamber to provide bees with space to store excess honey. Beekeepers must closely monitor honey production and add supers as needed to prevent overcrowding and encourage continued honey gathering. In fall, beekeepers assess colony strength, health, and food stores to ensure survival through the winter months. Inspections focus on evaluating honey reserves, reducing hive entrances to prevent robbing, and treating for pests and diseases. Beekeepers may also consider feeding colonies supplemental sugar syrup or pollen patties to bolster food stores and stimulate brood rearing before winter. Varroa mites pose a significant threat to bee colonies, particularly in the fall when populations peak. Beekeepers implement mite control measures such as using miticides, drone brood removal, and integrated pest management techniques to minimize winter losses. Monitoring mite levels regularly and treating colonies as needed is essential for maintaining healthy bee populations.

During winter, beekeepers may insulate hives to help colonies maintain warmth and conserve energy. Insulation options include wrapping hives with insulating materials such as foam or tar paper, providing upper ventilation to prevent moisture buildup, and using insulated hive covers or quilts to retain heat. Periodic checks on honey reserves are crucial during winter to prevent starvation. Beekeepers may use techniques such as hefting hives, tapping on frames to assess honey stores, or using an infrared camera to monitor cluster position and activity. If food reserves are low, beekeepers may supplement colonies with emergency feeding to ensure survival until spring.

Sustainable Hive Inspection Techniques

Minimizing disruption during hive inspections is essential for maintaining colony health and productivity while reducing stress on the bees. By choosing optimal times for inspections, handling bees gently, honing observation skills, and implementing integrated pest management (IPM) techniques, beekeepers can minimize disturbance and promote a harmonious relationship with their colonies. Selecting the right time for hive inspections is crucial for minimizing disruption. Bees are less likely to be disturbed during periods of calm weather, preferably on mild, sunny days when foragers are out collecting nectar and pollen. Avoid conducting inspections during extreme weather conditions, such as high winds, heavy rain, or extreme temperatures, as these conditions can agitate bees and increase the risk of defensive behavior. Handle bees with care, using slow, deliberate movements to minimize agitation and avoid crushing bees. Move calmly and smoothly around the hive, avoiding sudden gestures or loud noises that may startle the bees. Use smoke judiciously to calm the colony and reduce defensive responses, but avoid excessive or prolonged use, which can have the opposite effect. Develop keen observation skills to identify healthy brood patterns and detect signs of disease or pests during inspections. Healthy brood patterns exhibit uniformity, with larvae of consistent size and color arranged in tightly packed cells. Look for abnormalities such as sunken, discolored, or perforated brood cells, which may indicate diseases such as American foulbrood or pest infestations. Estimate colony strength by assessing the size and activity level of the colony during inspections. A strong, healthy colony will have a robust population of worker bees actively foraging, caring for brood, and maintaining hive cleanliness. Monitor population trends over time to detect changes or abnormalities that may signal health issues or environmental stressors.

Conduct regular checks for pests and diseases, including Varroa mites, hive beetles, foulbrood, and other common issues. Use techniques such as sugar roll or alcohol washes to monitor Varroa mite levels, inspect frames for signs of beetle infestation, and visually inspect brood patterns for symptoms of foulbrood diseases. Implement IPM strategies to manage pests and diseases without relying solely on chemical treatments. Non-chemical control methods include using screened bottom boards to trap and remove Varroa mites, practicing drone brood removal to interrupt mite reproduction cycles, and employing natural treatments such as essential oils or organic acids to deter pests. By integrating these methods into a comprehensive pest management plan, beekeepers can reduce reliance on chemical pesticides and minimize environmental impact while maintaining healthy bee populations.

CHAPTER 5

BEE HEALTH AND WELLNESS

Identifying and Preventing Common Bee Diseases

Identifying and preventing common bee diseases is crucial for maintaining healthy and productive colonies. By recognizing the signs of diseases such as varroosis, American foulbrood (AFB), nosemosis, and chalkbrood, and implementing preventative measures such as hygienic practices, genetic selection, and biosecurity protocols, beekeepers can safeguard their colonies against disease outbreaks.

Varroosis is caused by the parasitic Varroa mite, which weakens bees by feeding on their hemolymph and transmitting viruses. Signs of varroosis include deformed wings, crawling or weakened bees, and the presence of mites on adult bees and in brood cells during inspections. AFB is a bacterial disease caused by Paenibacillus larvae, which affects bee brood and can decimate entire colonies if left unchecked. Symptoms of AFB include a foul odor resembling rotting fish or sour milk, discolored larvae that darken and become ropy, and sunken, perforated cappings on affected cells. Nosemosis is another disease caused by the microsporidian parasite Nosema apis or Nosema ceranae, which infects the midgut of adult bees. Signs of nosemosis may include dysentery, where bees exhibit diarrhea-like symptoms, reduced colony strength, and decreased lifespan of worker bees due to compromised immune function. Chalkbrood is a fungal disease caused by the fungus Ascosphaera apis, which infects bee brood and mummifies larvae. Symptoms of chalkbrood include white, mummified larvae that resemble small, hard pellets and chalk-like spores present on affected brood comb.

Preventative Measures

Maintain clean equipment, minimize hive stress, and practice good apiary hygiene to reduce the risk of disease transmission. Regularly clean and sterilize hive components, replace old comb, and ensure proper ventilation and hive spacing to prevent moisture buildup and disease proliferation. Choose bee stocks with resistance or tolerance to common diseases when selecting queens or purchasing bees. Breeding for disease resistance is essential for building resilient colonies capable of withstanding disease pressure without the need for chemical treatments. Implement biosecurity protocols to prevent the introduction and spread of pathogens into the apiary. Quarantine new bees and equipment before introducing them into established colonies, and practice strict hygiene measures such as washing hands and tools between hive inspections to minimize disease transmission. By promptly recognizing the signs of common bee diseases and implementing preventative measures such as hygienic practices, genetic selection, and biosecurity protocols, beekeepers can protect their colonies from disease outbreaks and promote the health and longevity of their bee populations.

Natural Remedies and Treatments for Bee Health

Integrated Pest Management (IPM) is an approach to pest control that emphasizes the use of multiple strategies to manage pests while minimizing reliance on chemical treatments. By integrating mechanical, biological, herbal, and essential oil treatments, as well as probiotic and prebiotic supplements, beekeepers can effectively manage pest populations while promoting the health and resilience of their colonies.

1. **Mechanical Control:**
- **Drone Brood Removal:** Removing drone brood from the hive interrupts the reproductive cycle of Varroa mites, as they preferentially infest drone cells for reproduction. Beekeepers can use specialized frames or traps to collect drone brood, reducing mite populations within the colony.

- **Screened Bottom Boards:** Screened bottom boards provide a physical barrier to mites, preventing them from re-entering the hive after they fall off adult bees. Mites dislodged from grooming or natural behaviors fall through the screen and onto a removable tray, where they can be monitored and managed.

2. **Biological Control:**

- **Introduction of Beneficial Organisms:** Introducing predatory mites or fungi that target pest populations can help control pest infestations within the hive. Predatory mites such as Stratiolaelaps scimitus prey on Varroa mites, while fungi like Metarhizium anisopliae can infect and kill Varroa mites without harming bees.

3. **Herbal and Essential Oil Treatments:**

- **Thymol:** Thymol, derived from thyme oil, is a natural treatment for Varroa mites. It disrupts mite reproduction and respiration when applied as a vapor or in gel form within the hive. Thymol is typically applied during periods of broodlessness to minimize exposure to developing brood.

- **Formic Acid:** Formic acid is an organic acid that is effective against Varroa mites when applied as a fumigant or in gel pads within the hive. It disrupts mite respiration and reproductive cycles, leading to mite mortality. Formic acid treatments should be carefully monitored to prevent harm to bees and queen.

- **Essential Oils:** Essential oils such as tea tree, lemongrass, and peppermint have shown potential for controlling pests and diseases in bee colonies. These oils may have antimicrobial and insecticidal properties when applied in appropriate concentrations and formulations. However, caution should be exercised to avoid adverse effects on bees and hive health.

4. **Probiotic and Prebiotic Supplements:**
- **Probiotics:** Probiotics introduce beneficial bacteria into the bee gut microbiome, supporting gut health and enhancing disease resistance. Probiotic supplements may include lactic acid bacteria or other beneficial microorganisms that promote digestion and nutrient absorption in bees. Prebiotics provide

nutrient sources that selectively promote the growth of beneficial gut microbes in bees. These supplements may include oligosaccharides or other compounds that serve as food for beneficial bacteria, helping to maintain a healthy balance of gut microbiota in the hive.

By incorporating a diverse range of IPM strategies, beekeepers can effectively manage pest populations while minimizing the use of chemical treatments and promoting the health and resilience of their colonies. Regular monitoring, careful application of treatments, and ongoing research into new methods are essential components of successful pest management in beekeeping.

Monitoring and Maintaining Hive Wellness

Regular inspections are essential for maintaining healthy and productive bee colonies. By establishing a schedule for hive inspections, conducting comprehensive assessments, maintaining detailed records, and using diagnostic tools, beekeepers can monitor colony health, address issues promptly, and make informed management decisions. Develop a regular schedule for hive inspections based on seasonal needs and colony health. In the active foraging season, inspections may be conducted weekly or biweekly to monitor brood development, honey production, and pest activity. During periods of low activity or inclement weather, inspections may be less frequent but should still occur regularly to ensure colony vitality. Assess brood patterns during inspections to gauge colony health and productivity. Healthy brood patterns exhibit uniformity, with larvae of consistent size and color arranged in tightly packed cells. Abnormalities such as spotty brood or irregular cell cappings may indicate issues such as pest infestations, diseases, or nutritional deficiencies. Evaluate honey stores during inspections to ensure colonies have an adequate food supply for survival and growth. Monitor honey reserves by lifting hive boxes or gently tapping frames to assess weight. Supplemental feeding may be necessary if colonies are low on honey stores, especially during periods of nectar dearth or when preparing for winter. Observe colony population dynamics during inspections to assess overall strength and activity level. A healthy colony will have a robust population of worker bees actively foraging, caring for brood, and maintaining hive cleanliness. Monitor population trends over time to detect changes or abnormalities that may indicate health issues or environmental stressors. Maintain detailed records of hive observations, treatments, and conditions to track colony health over time. Use a hive inspection logbook or digital record-keeping system to document dates, observations, and actions taken during inspections. Health logs provide valuable insights into colony performance, disease trends, and treatment efficacy. Keep records of dates and types of treatments administered for pest and disease management. Note the specific products used, application methods, and outcomes of treatments.

Tracking treatment history allows beekeepers to monitor the effectiveness of interventions, identify recurring issues, and adjust management strategies as needed. Use techniques such as sugar shakes or alcohol washes to estimate Varroa mite infestation levels in bee colonies. Collect samples of adult bees and conduct mite counts to assess the severity of infestations and determine the need for treatment. Regular monitoring helps beekeepers implement timely interventions to control mite populations and minimize colony losses. When necessary, send samples of bees, brood, or hive materials to specialized laboratories for disease diagnosis and resistance testing. Laboratory analysis can identify specific pathogens, assess disease prevalence, and determine the presence of resistant strains. Test results provide valuable information for implementing targeted management strategies and selecting appropriate treatment options. Regular inspections, comprehensive assessments, diligent record-keeping, and judicious use of diagnostic tools are essential components of proactive hive management in beekeeping. By monitoring colony health and addressing issues promptly, beekeepers can promote the well-being and productivity of their bees and contribute to sustainable beekeeping practices.

CHAPTER 6

ENHANCING HONEY PRODUCTION SUSTAINABLY

Understanding Honey Flow and Forage

Understanding honey flow and forage is crucial for successful beekeeping, as it directly influences honey production and colony health. By comprehending the dynamics of honey flow, recognizing seasonal variations, and identifying key forage sources, beekeepers can optimize their management practices and ensure their bees have access to adequate nutrition.

Honey flow refers to the period when nectar-producing plants are in bloom and bees actively collect nectar to produce honey. This period is significant for beekeepers because it determines the availability of resources necessary for honey production and colony growth. Honey flow can vary greatly depending on the region, climate, and types of plants available. Honey flow periods vary throughout the year, typically aligning with the bloom cycles of major nectar sources. In many regions, the primary honey flow occurs in spring and early summer when a wide variety of plants are in bloom. A secondary flow may occur in late summer or early fall, depending on the presence of late-blooming plants. Understanding these seasonal variations helps beekeepers prepare for peak nectar collection periods and manage their colonies accordingly. Several factors influence honey flow, including weather conditions, plant bloom periods, and regional variations. Weather plays a crucial role, as adequate rainfall and favorable temperatures promote abundant nectar production, while extreme weather can reduce it. Plant bloom periods depend on the types of flora present in the area, with different plants blooming at various times of the year. Regional variations in climate and vegetation types also affect the timing and duration of honey flow.

Forage Sources

Key forage sources for bees include a variety of trees, flowers, and crops that produce nectar and pollen. Common nectar sources include flowering trees such as linden and maple, wildflowers like clover and dandelion, and cultivated crops such as sunflowers and canola. Pollen sources are equally important for providing the protein needed for brood rearing and colony development. Creating and utilizing plant bloom calendars helps beekeepers anticipate forage availability throughout the year. A bloom calendar lists the flowering periods of major nectar and pollen sources in a specific region, allowing beekeepers to track when their bees will have access to different plants. This tool is invaluable for planning hive management activities, such as adding honey supers or preparing for nectar dearth periods. Evaluating the quality and abundance of forage in the area is essential for supporting honey production and overall colony health. High-quality forage provides a rich supply of nectar and pollen, promoting robust honey production and strong colony growth. Beekeepers should assess their local environment to ensure there is a diverse range of flowering plants throughout the seasons. If natural forage is limited, beekeepers can enhance forage availability by planting bee-friendly gardens, wildflower meadows, or cover crops. Understanding honey flow and forage dynamics enables beekeepers to optimize hive management, ensure their colonies have access to sufficient resources, and maximize honey production. By staying attuned to seasonal variations, monitoring weather patterns, and identifying key forage sources, beekeepers can effectively support their bees' nutritional needs and contribute to the overall success and sustainability of their apiary.

Techniques to Boost Honey Yield Naturally

Effective hive management practices are essential for maintaining healthy colonies and maximizing honey production. By implementing strategies such as timely supering, efficient comb management, and maintaining colony strength, beekeepers can ensure their bees have the resources they need to thrive. Enhancing forage availability through planting for pollinators, integrating crops, and restoring habitats further supports bee health and productivity. Additionally, providing supplemental feeding, pollen traps, and water ensures that bees have adequate nutrition and hydration, particularly during challenging periods.

1. **Hive Management Practices**

- **Supering Strategy:** Adding supers at the right time is crucial for giving bees the space they need to store honey and preventing swarming. Supers should be added when the colony's population is strong and nectar flow is abundant. This encourages bees to move up and store honey in the new space, reducing congestion in the brood chamber. Proper timing of supering can also help manage the colony's growth and minimize the risk of swarming, which occurs when bees feel overcrowded and decide to split the colony.

- **Comb Management:** Encouraging bees to build and maintain comb efficiently is vital for optimal honey storage and brood rearing. Beekeepers can manage comb by rotating old frames out of the hive and replacing them with new foundation or drawn comb. This practice not only ensures that the comb remains in good condition but also helps control diseases and pests that can accumulate in old wax. Additionally, providing ample space and proper frame spacing can promote even comb building and prevent the formation of irregular or misshapen comb.

- **Colony Strength:** Maintaining strong, healthy colonies with adequate populations is key to maximizing foraging and honey production. Beekeepers should regularly monitor colony health and address any issues promptly, such as treating for pests and diseases, requeening if necessary, and ensuring adequate food stores. Strong colonies are better equipped to forage efficiently, defend against pests, and endure environmental stresses.

2. **Enhancing Forage Availability:**
- Planting for Pollinators: Establishing diverse plantings of nectar and pollen-rich species around the apiary provides bees with a

consistent and varied food source. Beekeepers can plant a mix of wildflowers, flowering shrubs, and trees that bloom at different times of the year to ensure continuous forage availability. Native plants are particularly beneficial as they are well adapted to the local environment and are often preferred by native pollinators.

- **Crop Integration:** Collaborating with local farmers to plant bee-friendly cover crops and forage plants can significantly enhance forage availability. Cover crops such as clover, alfalfa, and buckwheat not only improve soil health but also provide valuable nectar and pollen sources for bees. By working together, beekeepers and farmers can create mutually beneficial landscapes that support agricultural productivity and bee health.

- **Habitat Restoration:** Participating in or supporting initiatives to restore natural habitats and improve forage diversity is another effective strategy. Restoration projects may involve planting native vegetation, creating pollinator corridors, and removing invasive species that compete with native plants. These efforts help rebuild ecosystems that support a wide range of pollinators, including honeybees, and contribute to biodiversity conservation.

3. **Bee Health and Nutrition:**
- Supplemental Feeding: Providing supplemental feeding during nectar dearths is essential for maintaining colony strength

without over-reliance on stored honey. Sugar syrup can be provided as a carbohydrate source, while protein supplements or pollen substitutes can be offered to support brood rearing. It's important to monitor colony needs and adjust feeding practices accordingly to avoid overfeeding and potential issues such as dysentery.

- **Pollen Traps and Supplements:** Using pollen traps sparingly and providing high-quality pollen substitutes when necessary helps ensure bees have adequate protein for brood development and overall health. Pollen traps should be used judiciously to avoid depleting the colony's protein reserves. When natural pollen is scarce, beekeepers can offer commercial or homemade pollen substitutes to support colony nutrition.

- **Water Provision:** Ensuring bees have constant access to clean water sources is crucial, especially during hot and dry periods. Bees use water for cooling the hive, diluting honey, and maintaining hydration. Beekeepers can provide water by setting up shallow containers with floating materials (such as corks or stones) to prevent drowning. Regularly checking and replenishing water sources helps ensure bees stay healthy and productive.

By integrating these hive management practices, enhancing forage availability, and ensuring adequate bee health and nutrition, beekeepers can create an environment where their colonies can thrive. These strategies not only support honey production but also contribute to the overall resilience and sustainability of the apiary.

Harvesting Honey with Minimal Impact

Identifying the best times to harvest honey involves monitoring hive conditions and understanding seasonal factors. The optimal time for harvesting is typically late spring through early fall when nectar flow is at its peak. It's important to ensure that the bees have had enough time to produce and cap the honey. Beekeepers often wait until most of the frames in the honey supers are filled and capped with wax, indicating that the honey is ripe and ready for extraction. Ensuring honey is fully capped and ripened is crucial to prevent fermentation and maintain quality. Honey that is not fully ripened contains higher moisture content, which can lead to fermentation and spoilage. Beekeepers should inspect the frames and ensure that at least 80-90% of the honey cells are capped before harvesting. Using a refractometer to measure moisture content can also help ensure that the honey is at the ideal moisture level of 18% or less. Using bee-friendly tools and techniques helps minimize disturbance and stress during harvesting. Beekeepers can employ methods such as using a bee brush or a soft bee blower to gently remove bees from the frames. Honey harvesters should avoid rough handling of the frames and use smoke sparingly to calm the bees. Additionally, working during cooler parts of the day when bees are less active can reduce agitation.

Sustainable Harvest Practices

Leaving enough honey for bees to sustain themselves is especially important heading into winter. Beekeepers should practice partial harvesting, taking only the surplus honey and ensuring that the colony has enough reserves to survive periods of low nectar flow and cold weather. Typically, leaving 60-80 pounds of honey for the bees is recommended, depending on the climate and length of winter. Rotating and replacing frames judiciously helps maintain hive health and productivity. Old and damaged frames should be removed and replaced with fresh foundation or drawn comb. This practice helps prevent disease buildup and encourages bees to maintain clean and productive comb. Regularly inspecting and rotating frames also allows for better management of brood and honey stores. Utilizing beeswax sustainably involves recycling wax comb and considering the environmental impact of wax processing. Beekeepers can melt down old comb and purify the wax for use in candles, cosmetics, and other products. Reusing wax reduces waste and supports a circular economy within the apiary. Environmentally-friendly methods such as solar wax melters can be employed to process wax efficiently.

Post-Harvest Care

Proper techniques for storing harvested honey are essential to maintain quality and prevent spoilage. Honey should be stored in airtight, food-grade containers to protect it from moisture and contaminants. Keeping honey in a cool, dark place helps preserve its natural properties and prevent crystallization. Regularly checking stored honey for signs of fermentation or spoilage ensures its long-term quality. Accurately labeling honey with source information and maintaining records for traceability is important for both legal compliance and consumer trust. Labels should include details such as the type of honey, harvest date, and location of the apiary. Maintaining detailed records allows beekeepers to track the provenance of their honey and address any quality control issues. Choosing eco-friendly packaging materials helps minimize environmental impact. Beekeepers can opt for recyclable or biodegradable packaging options such as glass jars, paper labels, and cardboard boxes. Reducing the use of plastics and incorporating sustainable materials aligns with environmentally conscious practices and appeals to eco-minded consumers.

CHAPTER 7

PROCESSING AND SELLING HONEY

Sustainable Honey Processing Methods

1. **Manual vs. Electric Extractors:** Comparing the environmental impact, efficiency, and suitability for different scales of operation is crucial in choosing the right equipment. Manual extractors, which use hand-cranking mechanisms, are eco-friendlier as they do not require electricity and are suitable for small-scale operations. They are also quieter and produce less heat, preserving

the honey's natural qualities. Electric extractors, on the other hand, are more efficient for larger-scale operations, reducing labor time significantly. However, they consume electricity, and their environmental impact depends on the energy source. Solar-powered electric extractors can mitigate some of this impact.

2. **Cold Extraction Methods:** Using cold extraction methods offers benefits in preserving honey's natural enzymes and nutrients. Cold extraction avoids heating honey, which can degrade its beneficial properties and alter its flavor. Techniques like cold pressing and letting honey drip naturally from the comb into collection containers ensure that the final product retains its full

nutritional profile. This method is energy-efficient and aligns well with sustainable practices.

3. **Crush and Strain Method:** This simple, low-tech method is suitable for small-scale beekeepers and involves crushing the honeycomb to release honey, then straining it to remove wax and debris. The process requires minimal equipment and no electricity, making it an environmentally friendly option. However, it destroys the comb, requiring bees to rebuild it, which can be labor-intensive for the colony.

4. **Filtering and Purification:** Techniques for removing impurities while retaining pollen and natural elements include using coarse filters that allow small particles like pollen to

pass through while capturing larger debris. This method maintains the honey's natural texture and health benefits. Minimal filtration is less energy-intensive and preserves the ecological integrity of the honey. Utilizing gravity to filter honey naturally without additional energy input involves letting honey pass through filters slowly under its own weight. This method is energy-efficient and gentle on honey, preserving its enzymes and flavor. Gravity filtering can be done using layered filters of varying coarseness to achieve the desired clarity without over-processing. Ensuring minimal processing to maintain honey's natural qualities and reduce energy consumption involves limiting steps like excessive heating, fine filtration, and high-speed processing. By keeping the process simple and natural, beekeepers can offer a product that is closer to what bees produce, retaining its authenticity and health benefits.

5. **Sustainable Facility Management:** Implementing energy-efficient practices in honey processing facilities can significantly reduce the environmental footprint. This includes using solar power for electricity, installing energy-efficient appliances, and optimizing lighting and ventilation systems. Energy audits can help identify areas for improvement, and investments in renewable energy sources can further enhance sustainability. Techniques to minimize water use during cleaning and processing include using low-flow nozzles, recycling water where

possible, and implementing dry cleaning methods for equipment. Collecting and reusing rinse water for initial washes or non-critical cleaning tasks can also conserve water. Additionally, maintaining a clean working environment reduces the need for frequent deep cleaning. Strategies for reducing waste and recycling by-products like beeswax and propolis include implementing zero-waste practices and finding uses for all by-products. Beeswax can be purified and used for candles, cosmetics, and food wraps, while propolis can be processed into tinctures and health products. Composting organic waste and using biodegradable materials for packaging also contribute to waste reduction. Educating staff and adopting a circular economy mindset ensures that resources are used efficiently and sustainably.

By focusing on sustainable honey processing methods, beekeepers can enhance the quality of their honey, reduce environmental impact, and support the health of their bees and ecosystems. These practices not only contribute to a more sustainable apiary but also appeal to environmentally conscious consumers.

Packaging and Labeling for Eco-Friendly Sales

Eco-Friendly Packaging Options

1. **Glass Jars:** Using glass jars for honey packaging offers several benefits. Glass is highly recyclable and can be reused multiple times without losing quality, making it a sustainable choice. It also has a premium feel,

which can enhance consumer perception of the product's quality. Glass jars are inert, meaning they do not react with honey and preserve its flavor and purity. However, challenges include the heavier weight of glass, which increases transportation costs and carbon footprint. Additionally, glass is more prone to breakage, requiring careful handling and potentially higher packaging costs to ensure safe delivery.

2. **Biodegradable Packaging:** Exploring biodegradable and compostable packaging alternatives is essential for reducing plastic waste. Options include containers made from plant-based materials, such as PLA (polylactic acid) derived from cornstarch, or paper-based containers coated with beeswax or other natural substances. These materials break

down more quickly in the environment compared to conventional plastics, reducing pollution. Challenges include ensuring these materials provide adequate barrier properties to protect honey and maintain its shelf life. Also, consumer education is necessary to ensure proper disposal and composting of biodegradable packaging.

3. **Reusable Containers:** Encouraging the use of reusable containers can be achieved through deposit schemes or discounts. Customers can return empty containers for a refund or a discount on their next purchase, incentivizing them to participate in a circular economy. Reusable containers, such as sturdy glass jars or metal tins, can be washed and refilled, significantly reducing waste. Implementing

this system requires logistical planning to manage returns and ensure containers are sanitized properly before reuse.

Labeling Requirements and Best Practices

1. **Regulatory Compliance:** Ensuring labels meet local and national regulations is crucial for legal compliance and consumer safety. Labels must accurately reflect the product's content, including any additives or processing methods. Health claims must be substantiated and adhere to regulatory standards to avoid misleading consumers. Including nutritional information, origin, and any certifications, such as organic or fair trade, helps meet regulatory requirements and consumer expectations.

2. **Transparent Labeling:** Clearly communicating honey origin, floral sources, and production methods builds consumer trust and transparency. Labels should specify the region where the honey was harvested and the primary floral sources if known. Descriptions of beekeeping practices, such as organic methods or sustainable harvesting, provide additional assurance to consumers seeking ethically produced honey. Transparent labeling fosters loyalty and can differentiate products in a competitive market.

3. **Sustainability Claims:** Using labels to highlight sustainable practices, such as organic certification or eco-friendly packaging, appeals to eco-conscious

consumers. Certifications from recognized organizations, such as USDA Organic or Fair Trade, provide third-party validation of sustainability claims. Labels can also emphasize environmentally friendly practices, like using renewable energy in production or supporting local biodiversity through planting bee-friendly flora. Clearly presenting these claims helps attract consumers who prioritize sustainability in their purchasing decisions.

Design Considerations

1. **Brand Identity:** Creating a brand identity that reflects sustainability values is key to appealing to eco-conscious consumers. This involves using design elements that convey environmental responsibility, such as earthy color palettes, natural imagery, and simple,

elegant designs that suggest purity and quality. Consistency in branding across all touchpoints, from packaging to marketing materials, reinforces the brand's commitment to sustainability and helps build a strong, recognizable identity.

2. **Informative Labels:** Including information on bee health, environmental impact, and ways consumers can support sustainable beekeeping educates and engages customers. Labels can feature brief stories or facts about the importance of bees to ecosystems, the challenges they face, and the role of sustainable beekeeping in protecting bee populations. Providing tips for consumers, such as planting bee-friendly flowers or avoiding pesticides, encourages them to

contribute to environmental efforts. Informative labels not only enhance the product's appeal but also foster a connection between consumers and the broader mission of sustainability.

By integrating eco-friendly packaging options, complying with labeling regulations, and thoughtfully designing labels, beekeepers can effectively communicate their commitment to sustainability. These practices not only reduce environmental impact but also resonate with consumers who value ethical and environmentally responsible products.

CHAPTER 8

BREEDING AND RAISING QUEENS

Sustainable Queen Rearing Practices

The queen bee plays a crucial role in maintaining colony health, productivity, and genetic diversity. As the sole egg-layer, the queen determines the population size and growth rate of the colony. Her pheromones regulate hive behavior and cohesion, ensuring smooth operation and cooperation among worker bees. The genetic traits of the queen also influence the colony's resilience to diseases, temperament, and efficiency in foraging. Optimal times for raising queens vary based on regional climate and forage availability. Generally, spring and early summer are the best times for queen rearing because colonies are expanding, and there is an abundance of nectar and pollen. This period aligns with the natural swarming season when bees naturally rear new queens. Regional variations in climate and floral cycles must be considered to ensure queens are raised when conditions are favorable.

Queen Rearing Methods

1. **Grafting:** Grafting involves transferring young larvae from worker cells into specially

designed queen cups. This method requires precision and skill, using tools such as grafting needles. Best practices include selecting larvae that are less than 24 hours old and ensuring they are kept moist during the transfer. This method allows beekeepers to produce a large number of queens simultaneously and control the genetics of the new queens by choosing larvae from selected breeder colonies.

2. **Non-Grafting Methods:** Alternatives like the Jenter or Nicot systems simplify queen rearing by eliminating the need to manually transfer larvae. These systems involve placing a queen in a specially designed cage where she lays eggs in pre-formed cells. The eggs hatch into larvae, which can then be transferred directly into queenless colonies to be raised as

queens. These methods are user-friendly and reduce the skill level required, making queen rearing accessible to more beekeepers.

3. **Natural Queen Rearing:** Encouraging bees to raise queens naturally involves creating conditions that stimulate queen cell development. This can be achieved by splitting strong colonies or creating artificial swarming scenarios where the bees sense the need to produce a new queen. By reducing interventions, this method relies on the bees' natural instincts and minimizes stress, promoting healthy queen development.

Sustainable Practices

Reducing hive disturbance and handling stress is critical during queen rearing. Gentle handling techniques, minimal hive inspections, and maintaining stable hive environments help reduce stress on the bees. Stress can negatively impact the quality and acceptance of the new queens, so creating a calm and stable environment is essential. Ensuring that colonies raising queens have adequate resources is vital for their success. This includes providing ample food supplies, such as sugar syrup and pollen supplements if natural forage is insufficient. Sufficient space must be available within the hive to accommodate the growth and activities of the queen-rearing process, preventing overcrowding and ensuring the bees can focus on rearing healthy queens. Using strong, disease-free colonies as breeder hives is fundamental to promoting healthy queen development. Selecting colonies with desirable traits, such as good temperament, high productivity, and disease resistance, ensures that these traits are passed on to the new queens. Regular health checks and maintaining good apiary hygiene practices help prevent the spread of diseases and pests, supporting the development of robust queens. By adopting sustainable queen rearing practices, beekeepers can produce high-quality queens that enhance the health and productivity of their colonies. These methods not only contribute to the long-term sustainability of the beekeeping operation but also support broader environmental and ecological health by promoting genetic diversity and resilience in bee populations.

Selecting and Breeding for Desirable Traits
Key Traits for Selection

1. **Disease Resistance:** Identifying and selecting bees that exhibit natural resistance to common diseases and pests is crucial for sustainable beekeeping. This involves monitoring colonies for signs of diseases like American Foulbrood, Nosema, and pests like Varroa mites. Bees that show lower incidences of these issues or that survive better under high disease pressure can be used as breeding stock. This trait reduces the need for chemical treatments, promoting healthier hives and a more environmentally friendly apiary.

2. **Productivity:** Breeding for high honey production, pollen collection, and efficient foraging behaviors ensures that colonies maximize their resource intake. Productive colonies not only generate more honey but also

contribute to the overall health of the hive by ensuring there is enough pollen and nectar to support brood rearing and winter stores. This involves selecting queens from colonies that consistently produce high yields and demonstrate strong foraging activity.

3. **Gentle Temperament:** Selecting for docility makes hive management safer and more enjoyable. Gentle bees are less likely to sting, which reduces stress for both the bees and the beekeeper. This trait is particularly important for urban beekeeping or beekeepers who frequently interact with their hives. Colonies are evaluated for their behavior during routine inspections, with preference given to those that remain calm and manageable.

4. **Overwintering Success:** Prioritizing traits that enhance colony survival during winter months is essential for long-term beekeeping success. This includes selecting bees that effectively store honey and pollen, cluster tightly to conserve warmth, and display strong hygienic behaviors to prevent disease buildup during the winter. Colonies that consistently survive and thrive through harsh winters are excellent candidates for breeding.

Breeding Techniques

1. **Controlled Mating:** Controlled mating involves using isolated mating yards or instrumental insemination to ensure specific genetic pairings. In isolated mating yards,

queens are mated with drones from selected colonies to maintain desired traits. Instrumental insemination allows for precise control over the mating process by manually inseminating queens with sperm from selected drones. This technique ensures the purity of genetic lines and the consistency of desired traits.

2. **Open Mating:** Open mating allows queens to mate naturally in areas with high densities of desired drone genetics. This method is less labor-intensive than controlled mating and takes advantage of the natural mating behavior of bees. By maintaining drone colonies with favorable traits in proximity to mating yards, beekeepers can influence the genetic outcome

of open-mated queens, although with less precision than controlled methods.

3. **Selection Programs:** Implementing structured breeding programs involves tracking and selecting for desired traits over multiple generations. This systematic approach includes detailed record-keeping of colony performance, health, and behavior. By consistently selecting the best-performing colonies for breeding, beekeepers can incrementally improve the overall quality and resilience of their bee populations.

Genetic Diversity

1. **Maintaining Genetic Variation:** Ensuring a broad genetic base is essential to prevent

inbreeding and enhance colony resilience. Genetic diversity reduces the likelihood of hereditary diseases and improves the overall adaptability of bees to changing environmental conditions. Beekeepers can maintain genetic variation by incorporating queens and drones from different genetic lines into their breeding programs.

2. **Cross-Breeding:** Introducing new genetic lines periodically helps increase diversity and introduce beneficial traits. This can involve purchasing queens or drone semen from other beekeepers or breeders who specialize in different strains of bees. Cross-breeding can introduce new strengths, such as improved disease resistance or productivity, and helps

mitigate the risks associated with a limited genetic pool.

By focusing on these key traits and employing advanced breeding techniques, beekeepers can develop robust, productive, and resilient bee populations. Sustainable breeding practices not only enhance the health and productivity of individual colonies but also contribute to the broader health of bee populations and the ecosystems they support.

Integrating New Queens into Your Apiary
Queen Introduction Methods

1. **Direct Introduction:** Direct introduction involves placing the new queen directly into the hive using techniques to minimize rejection. This method can be suitable for

strong, queenless colonies that are in urgent need of a new queen. Care must be taken to ensure the queen is introduced gently and without causing undue disturbance to the colony.

2. **Indirect Introduction:** Indirect introduction methods involve using queen cages or push-in cages to allow the colony to acclimate to the new queen's pheromones gradually. These cages typically contain the queen and a small number of attendant worker bees. The cages are placed within the hive, allowing the workers to become accustomed to the queen's scent before her release.

3. **Nucleus Colony Introduction:** Introducing the new queen into a smaller nucleus colony before merging with a larger

hive can increase her chances of acceptance. Nucleus colonies provide a controlled environment where the queen can establish herself and begin laying eggs before being introduced to the main hive. This method can be particularly useful for introducing queens with unknown genetics or in situations where acceptance may be uncertain.

Monitoring and Management

1. **Acceptance Signs:** Observing colony behavior is essential to ensure the new queen is accepted. Signs of acceptance include a lack of aggression towards the queen, workers attending to her, and the presence of brood laid by the queen. Queen-right behavior, such as the absence of queen cells, is another indicator of successful acceptance.

2. **Post-Introduction Care:** Providing additional resources and monitoring closely are necessary to support the new queen's establishment. This may include supplementary feeding to ensure the colony has ample resources during the transition period. Regular inspections allow beekeepers to assess the queen's laying pattern and overall health, ensuring she is fulfilling her role effectively.

3. **Addressing Rejection:** If the new queen is rejected, beekeepers must be prepared to intervene promptly. This may involve reintroducing the queen using a different method or replacing her with a new queen. Understanding the reasons for rejection, such as poor queen quality or disturbances during

introduction, can help inform future introduction strategies.

Record-Keeping and Evaluation

1. **Tracking Queen Lineage:** Keeping detailed records of queen genetics, traits, and performance is essential for informed breeding decisions. This includes information on the queen's lineage, breeding history, and any observed traits or behaviors. By tracking lineage, beekeepers can identify successful breeding lines and make informed choices about future queens.

2. **Performance Evaluation:** Assessing the new queen's impact on colony health, productivity, and behavior over time provides valuable insights into her effectiveness. This evaluation may include monitoring honey

production, brood patterns, temperament, and disease resistance. By systematically evaluating queen performance, beekeepers can identify queens that excel and incorporate their genetics into future breeding programs.

3. **Continuous Improvement:** Using data from evaluations to refine and improve breeding practices and selection criteria is essential for ongoing success. This may involve adjusting breeding goals based on observed performance, selecting for specific traits that contribute to colony resilience and productivity, and culling poorly performing queens from breeding programs. Continuous improvement ensures that breeding efforts remain focused on producing high-quality

queens that contribute positively to hive health and productivity.

CHAPTER 9

EXPANDING AND DIVERSIFYING YOUR APIARY

Increasing Hive Numbers Sustainably

Before increasing hive numbers, it's essential to evaluate the suitability of the current apiary site. Factors such as forage availability, space, and environmental conditions should be considered. Ensure that the site can support additional hives without causing overcrowding or resource shortages. Access to sufficient food sources, water, and shelter is critical for expanding colonies sustainably. Adequate forage availability throughout the year ensures that colonies have access to essential nutrients for growth and productivity. Additionally, providing sources of clean water nearby helps maintain colony hydration, especially during hot and dry periods.

Expansion Methods

1. **Splitting Colonies:** Splitting colonies involves dividing existing colonies to create new hives. Techniques such as the walk-away split or artificial swarm can be used to accomplish this. The walk-away split involves separating frames of brood, bees, and resources into new hive bodies, allowing the bees to raise a new queen in each split. Artificial swarms mimic the natural swarming process by creating a nucleus colony from an existing hive.

2. **Nucleus Colony Production:** Establishing nucleus colonies (nucs) is a sustainable way to increase hive numbers with minimal disruption to existing colonies. Nucs consist of a small number of frames containing brood,

bees, and a mated queen. They serve as miniature colonies that can be expanded into full-size hives over time. Nuc production allows beekeepers to raise queens, manage genetics, and provide starter colonies for new beekeepers.

3. **Swarm Management:** Capturing and hiving swarms can help expand apiary populations while preventing colony losses. Swarms are natural means of colony reproduction, and capturing them allows beekeepers to harness this reproductive potential. By providing suitable swarm traps or bait hives, beekeepers can attract and capture swarms before they establish themselves in undesirable locations, such as inside buildings or trees.

Expanding hive numbers sustainably requires careful consideration of both the apiary site's capacity and available resources. Employing methods such as colony splitting, nucleus colony production, and swarm management allows beekeepers to increase hive numbers while promoting colony health and productivity. Additionally, regular monitoring and management ensure that expansion occurs in a way that benefits both the beekeeper and the bees.

Exploring Alternative Apiary Products (beeswax, propolis, etc.)
Beeswax Production and Uses

Beeswax can be harvested during honey extraction by collecting wax cappings that bees use to seal honey cells. Alternatively, beekeepers can render wax from old comb that has been replaced during hive maintenance. The harvested wax is then processed to remove impurities and excess moisture. Beeswax is a versatile natural material used in various products. It's commonly found in candles, where its clean-burning properties and natural aroma make it an ideal choice. Additionally, beeswax is used in balms, creams, and cosmetics for its emollient and skin-nourishing properties. Beekeepers can add value to beeswax by incorporating it into handmade products through molding, scenting, or combining it with other natural ingredients such as essential oils or herbs. This allows for the creation of unique and artisanal items that appeal to consumers seeking natural and sustainable alternatives.

Propolis Harvesting and Applications

Propolis, also known as bee glue, is a resinous substance collected by bees from tree sap and other botanical sources. Beekeepers can gather propolis from hive surfaces or by using traps placed within the hive. Propolis traps encourage bees to deposit excess propolis, which can then be collected and processed. Propolis has a long history of medicinal use due to its antimicrobial, anti-inflammatory, and antioxidant properties. It's often used in tinctures, salves, and supplements for its potential health benefits, including immune support, wound healing, and oral health. In addition to its medicinal properties, propolis is valued in craft and artisanal applications. It can be incorporated into products such as varnishes, paints, and woodworking finishes for its natural adhesive properties and beautiful amber color. Propolis-infused products add a unique and sustainable touch to artisanal crafts and woodworking projects. Both beeswax and propolis offer a range of uses beyond their traditional roles in the hive. From skincare products to artisanal crafts, these natural bee products provide opportunities for beekeepers to diversify their product offerings and add value to their beekeeping enterprises.

Value-Added Products and Services
Honey Varietals and Infusions

Offering single-source honey varietals sourced from specific floral nectar sources provides consumers with a taste of the unique flavors and aromas of different floral varieties. Varietals such as lavender, orange blossom, or wildflower offer distinct characteristics based on the plants visited by the bees. Experimenting with infused honeys allows beekeepers to create unique flavor profiles by incorporating natural ingredients like herbs, spices, or fruits. Infused honeys offer a creative way to enhance the natural sweetness of honey with complementary flavors, appealing to consumers seeking innovative and gourmet products.

Pollination Services

Beekeepers can provide pollination services to local farmers and orchardists, helping to enhance crop yields through increased pollination rates. Commercial pollination contracts offer beekeepers an additional source of income while playing a crucial role in supporting agricultural production. Collaborating with urban gardens, community farms, and green spaces allows beekeepers to support pollinator-friendly habitats in urban environments. Urban pollination partnerships not only benefit local gardens and green spaces but also raise awareness about the importance of pollinators in urban ecosystems.

Educational Workshops and Tours

Offering beekeeping workshops provides aspiring beekeepers with the knowledge and skills needed to start and maintain their own hives. Beginner and advanced workshops cover topics such as hive management, pest and disease control, and honey harvesting, empowering participants to become successful beekeepers. Inviting the public to visit the apiary for guided tours offers an opportunity to educate individuals about bees, pollination, and sustainable agriculture. Apiary tours provide firsthand experience with beekeeping practices, hive dynamics, and the importance of bees in food production, fostering a deeper appreciation for the natural world.

Hive Products Experiences

Hosting events such as honey tastings, beeswax candle-making workshops, or propolis harvesting demonstrations allows consumers to engage with hive products in meaningful ways. Farm-to-table experiences highlight the connection between bees, agriculture, and the products they produce, promoting sustainability and local food systems.

CHAPTER 10

INTEGRATING BEEKEEPING WITH LOCAL ECOSYSTEMS

Enhancing Biodiversity in Your Apiary

Incorporating native flowering plants into apiary landscapes enriches forage diversity by providing bees with a variety of nectar and pollen sources adapted to the local ecosystem. Native plants are well-suited to the climate and soil conditions, supporting pollinator populations and enhancing overall biodiversity. Ensuring year-round forage availability is essential for maintaining healthy bee populations. Planting species that bloom at different times of the year extends the foraging season and provides bees with continuous access to food sources, reducing periods of scarcity and enhancing colony resilience. Preserving or establishing wooded areas and hedgerows near apiaries creates habitat diversity and enhances foraging opportunities for bees. These natural features provide nesting sites, shelter, and additional forage for native pollinators, contributing to overall ecosystem health and resilience. Creating or maintaining water sources such as ponds, streams, or birdbaths within or near apiary sites is essential for bee hydration and habitat diversity. Bees require access to clean water for drinking and cooling the hive, and providing suitable water sources supports their health and well-being. Minimizing or eliminating pesticide and herbicide use in and around apiary sites is crucial for protecting pollinators and their forage plants. Chemical inputs can have harmful effects on bee health and disrupt ecosystem balance, leading to declines in pollinator populations and biodiversity. Adopting mowing practices that allow flowering plants to bloom and provide forage for bees and other pollinators promotes habitat diversity and enhances apiary biodiversity. Leaving areas of grass or wildflowers uncut allows pollinator-friendly plants to flourish, supporting bee populations and contributing to a healthier ecosystem overall.

Planting for Pollinators: Creating Bee-Friendly Habitats

Choosing native plants is essential for supporting local pollinator populations. Native species are well-adapted to the region's soil and climate conditions, providing reliable nectar and pollen sources for bees throughout the year. By selecting native plants, beekeepers can promote biodiversity and ecosystem resilience. Opting for plants with high nectar and pollen yields ensures abundant forage for bees. Wildflowers, herbs, and fruit trees are excellent choices for pollinator gardens, as they offer a diverse range of floral resources and support a variety of pollinator species. Designing pollinator gardens involves planning layouts that provide shelter, nesting sites, and continuous blooms throughout the growing season. Incorporating a mix of flowering plants with varying bloom times ensures a steady supply of food for bees and other pollinators year-round. Practicing bee-safe gardening techniques is crucial for protecting pollinators from harm. Avoiding or minimizing the use of pesticides and herbicides in pollinator gardens helps maintain a healthy environment for bees and other beneficial insects. Instead, focus on natural pest management methods and integrated pest management practices. Installing bee hotels or nesting boxes provides essential nesting sites for solitary bees and other native pollinators. These structures mimic natural nesting environments, offering cavities for solitary bees to lay their eggs and rear their young. Bee hotels contribute to habitat enhancement and support pollinator diversity. Converting lawn areas into wildflower meadows or pollinator-friendly habitats is a powerful way to enhance habitat for bees. Meadows increase floral diversity and forage availability, providing bees with ample nectar and pollen sources. Meadow establishment helps restore biodiversity and create vibrant ecosystems that benefit both bees and humans.

Working with Local Farmers and Gardeners

Partnering with local farmers to sell honey and bee-related products at farmers' markets creates opportunities for mutual support and community engagement. By showcasing locally produced honey and educating consumers about the importance of pollinators, beekeepers can raise awareness and build relationships with both farmers and consumers. Establishing agreements with farmers to provide pollination services in exchange for access to diverse forage sources and land for apiary expansion benefits both parties. Beekeepers gain access to additional forage sources, while farmers benefit from increased crop yields through improved pollination services, fostering a symbiotic relationship between beekeepers and agricultural producers. Offering educational workshops and presentations to local farmers and gardeners provides valuable information on the importance of pollinators, sustainable beekeeping practices, and creating pollinator-friendly habitats. By sharing knowledge and best practices, beekeepers can empower others to take action in supporting pollinator health and habitat conservation.

Collaborating with local garden centers or agricultural extension offices to establish demonstration gardens showcases bee-friendly plants and habitat enhancement techniques in action. These gardens serve as practical examples for the community, inspiring individuals to incorporate pollinator-friendly practices into their own gardens and landscapes. Advocating for pollinator-friendly policies at the local level, such as restrictions on pesticide use in public spaces or incentives for pollinator habitat conservation, helps create supportive environments for bees and other pollinators. Beekeepers can engage with local governments and community organizations to promote policies that prioritize pollinator health and habitat protection. Supporting efforts to conserve and restore natural habitats through land trusts, conservation easements, or community-led initiatives preserves essential forage sources and nesting sites for pollinators. By participating in land conservation efforts, beekeepers contribute to broader conservation goals and help safeguard biodiversity for future generations.

CONCLUSION

In the journey through this comprehensive guide to bee farming, we have delved into the intricate world of beekeeping, exploring not only the practical aspects of hive management but also the profound relationship between bees and their environment. From the basics of bee biology to the intricacies of sustainable hive management, we have uncovered the key principles and practices that underpin successful and ethical beekeeping endeavors. At the heart of sustainable bee farming lies a profound respect for the intricate balance of nature and a commitment to stewardship of the land. As beekeepers, we are not merely caretakers of hives; we are custodians of ecosystems, guardians of biodiversity, and partners in pollination. By embracing sustainable practices, we have the power to transform our apiaries into vibrant hubs of life, where bees thrive, ecosystems flourish, and communities prosper. Throughout this guide, we have emphasized the importance of adopting holistic approaches to beekeeping that prioritize the well-being of bees, the health of the environment, and the sustainability of our practices. From enhancing biodiversity in our apiaries to forging partnerships with local farmers and gardeners, we have explored a myriad of strategies for integrating beekeeping with local ecosystems and fostering resilience in the face of environmental challenges. As we conclude our exploration, let us reaffirm our commitment to sustainable bee farming—a commitment to nurturing healthy colonies, fostering vibrant habitats, and safeguarding the invaluable services that bees provide to our planet. Whether you are a seasoned beekeeper or embarking on your first hive, may this guide serve as a beacon of inspiration and guidance on your journey toward sustainable bee farming. In the words of Rachel Carson, "In nature, nothing exists alone." As we embark on this collective endeavor to safeguard the future of bees and the ecosystems they inhabit, let us remember that our actions, however small, have the power to shape the world around us. Together, let us forge a future where bees thrive, biodiversity flourishes, and sustainable bee farming becomes not just a practice, but a way of life.

www.ingramcontent.com/pod-product-compliance
Lightning Source LLC
Chambersburg PA
CBHW052319220526
45472CB00001B/194